TRAGEDY IN GALT

The May 2nd, 1956 CP Rail Crash

PAUL LANGAN

ISBN - 978-0-9685867-7-8

Tragedy in Galt: The May 2nd,1956, CP Rail Crash

2nd edition (revised)

brentunespublishing@yahoo.ca

TABLE OF CONTENTS

INTRODUCTION

Around twenty years ago I published 5 local history booklets. One of these booklets, Tragedy in Galt - The May 2nd, 1956, CP Rail Crash, was very popular when it was released in 1999. Many parents and children vividly remembered visiting the crash site.

In the 1st edition, I was very fortunate to be able to interview a crew member Ronald Westworth and a first responder Len Mills that were there that fateful night.

In 2021, I decided to significantly revise and expand the book. Some of the additions are:

- 2002 interview with fireman William Hopkins
- 2006 newspaper interview with engineer George Nutkins who was seriously injured in the crash
- 2020 interview with brakeman Ken Shillington
- More newspaper articles from The Evening Reporter
- New crash scene photos

Special thanks to Dr. Debra Langan, editor, Ken Shillington for sharing his memories of that day and Dan Schmalz at the City of Cambridge Archives.

I hope you enjoy this revised and expanded version of the Tragedy in Galt - The May 2nd, 1956 CP Rail Crash

Paul Langan
Hespeler, Ontario
January 2021

THE ACCIDENT

On May 2nd, 1956 at 6:30pm, Canadian Pacific steam doubleheader "through" freight train #903 (with engines # 2205 & 5420) left Toronto with 40 cars. It was heading for Windsor. It was two hours and thirty minutes late. After arriving at Guelph Junction, #903 was delayed one hour further.

The train crew was Lead Locomotive Engineer William Palmer, Toronto; Fireman Henry Blocka, Toronto; Brakeman Tommy Watson; Galt, Second Locomotive Engineer George Nutkins, London; Fireman William Hopkins, London; and in the caboose were Brakeman Ronald Westworth, London and Conductor Jack Cowing, London.

At the Galt CPR station, extra or "slow" steam doubleheader train #5186 (with engines #5607 & 5126) was in the siding. The slow freight had dropped off 14 cars in Galt. The train crew on train #5186 was Lead Engine Engineer William Hunter, Toronto; Fireman A.H. Holden, Toronto; Brakeman R. Gates, London; Second Locomotive Engineer Tim Ready, London; Fireman Jack Easby, London; Brakeman Ken Shillington and Conductor Gerry Rivard, London.

Through freight #903 had priority on the main line over extra freight #5186. Although the crew of #5186 knew that #903 was late; they did not know the exact whereabouts of the through freight.

There were block signals at the east and west ends of the Galt yard. When through freight #903 entered the east end of the Galt yard the lights were green to proceed. Meanwhile train #5186 was in the siding. Then train #5186 moved onto the mainline unaware of the oncoming freight #903.

Meanwhile in the Galt station, Dispatcher James Stewart noticed that the indicator had gone down in the office. This signalled that freight #903 was in the yard. James ran out into the yard to warn the crew of #5186 to stay in the siding. Tragically his message was not understood. Through train #903 smashed into the back of box cars of train #5186. The impact of the crash caused the Canadian Pacific Railway overpass, a heavy steel-girder bridge, to collapse onto Water Street, the two engines of #903 plunging to the street.

Noted in The Evening Reporter newspaper article of May 3rd, 1956: "In the terrific crash, which shook windows for a block, the two engines of the through train crumpled through the bridge. Several box cars, with the tender, splintered, reared over the bridge. Steam hissed out, while a boiler exploded with a roar. Coal from the crushed freight car was scattered over the Water Street pavement as far south as Front Street."

"It was a terrible sight, one that a person would not forget in a hurry," said Fire Chief Walter Scott.

Inside the lead engine of #903 was engineer William Palmer. He sustained severe steam burns, was crushed and died instantly.

Brakeman Tom Watson from Galt was on his first paid run as a brakeman, and was also killed instantly. Tom and his parents and brother lived two miles west of the crash scene and their family house was located about 100 yards from the C.P. R. tracks. More information on the Watson family is in another chapter in his book.

Fireman Henry "Albert" Blocka of the lead engine jumped out of the engine and landed on the steps and rolled to the street. He was able to get up and escaped being crushed by the falling locomotives. He suffered broken ribs.

In the second engine of train #903 engineer, George Nutkins suffered critical injuries including broken wrists, a broken elbow, a broken back and extreme shock. The coroner asked his wife to come from London due to the severity of the injuries. Fortunately, George Nutkins survived the crash and went on to become a supervisor with the C.P.R. He died in his 88th year in 2011. A 2006 interview with George and London Free Press reporter April Kemick is in this book.

Firemen of the Second Engine on #903, William Hopkins, was hospitalized for shock and unknown injuries. An interview with William is in this book.

Ken Shillington, brakeman on the second locomotive of the "extra" Train #5186 was uninjured. Ken shares his memories of that day in this book.

Len Mills was a fireman for the Galt Fire Department and was one of the first people to arrive at the crash scene. His story is also located in this book.

The crash scene was besieged by hundreds of people as Galt citizens viewed the worst rail crash in their history. The carnage of twisted steel and steam from the locomotives was a sight few people would forget in their lifetime.

EYEWITNESSES TELL GRIM STORY

The following is an article by Claude Kewley of The Evening Reporters published May 3rd, 1956.

To the writer, it seems something akin to a miracle that eight of the 14 crewmen managed to escape injury in the train wreck here Wednesday night, when two were killed and four injured.

As I watched the rescue workers toiling in the midst of the mass of twisted steel, it seemed impossible the others could have escaped unhurt.

Somewhat tardy in arriving at the scene, by the time I got there, police had blockaded the entrance to Water Street at the Delta and cars were moving at a snail' space along the Dundas highway in consequence of ever-increasing traffic. People were still arriving by the hundreds after midnight and some stayed until the bodies of Wiliam Palmer, Toronto and Tommy Watson of the Blenheim Road had been removed.

Wreckers Arrive

The first of the wrecking crews arrived from London about midnight, accompanied by a handful of C.P.R. technicians. It was not until 3:30am that the special train pulling the big hook arrived at the scene.

Force of the collision knocked the east end of the steel bridge into Water Street and two of the locomotives straddled it, their noses buried in the concrete.

One C.P.R. official told Mayor Anderson he thought it might be necessary to pull both locomotives down to street level before repairs could commence. Meanwhile, Chicago-Montreal traffic is being rerouted.

Newsmen, radio and T-V people were attracted to the scene from various parts of the province. Noticed were representatives of the Kitchener and Hamilton T-V stations, a team of Toronto Star reporters and photographers, representative of the Toronto Telegram, The Hamilton Spectator, The London Free Press, the Kitchener Record, the Guelph Mercury and the Brantford Expositor.

Phoned Alarm

One of several who phoned the alarm to police and firemen was Mrs. Archie Waring, Samuelson Street, whose husband is a printer in the employ of The Evening Reporter.

"I happened to look out through the living room window," Mrs. Waring said, "and several freight cars lurching from the tracks. There was a grinding noise and a terrific thud at the moment of impact. Almost immediately clouds of steam filled the subway and curled high in the air."

The Waring residence is situated about 75 yards from the scene of the tragedy. "The doubleheader freight going west didn't appear to be travelling fast," she said.

"I've seen some grim sights and raw courage since I became an ambulance driver," said Amio Burkhardt who raced to the scene along with city ambulance, operated by Fraser's Taxi. "This fellow, whose name I don't know was lying on the sideway but when he saw us with the stretcher, he was more concerned with the others than himself. All the way to the hospital he mumbled about what happened to his buddies.

The body of Engineer Palmer was first discovered by Fire Department Captain Bob Murphy.

Crawling over and around the wrecked locomotives, Capt. Murphy discovered the body pinned between the nose of one locomotive and the cab of the other.

Assisting the others, Capt. Henry Johnson stumbled and fell backward and gashed one of his legs. He was treated at the scene and later received stitches at the hospital.

One of the first on scene along with the firemen was Sergeant Joe McCabe, who speedily organized traffic control with the aid of the firemen who cordoned off the area with ropes to restrain the crowd which grew into the thousands in a space of an hour.

Bodies Removed

In a state of shock Mrs. Alex Watson, mother of the youngest victim was treated by a physician who administered sedatives after which she went into a deep sleep. The father, Alex Watson, remained at the scene until after his son's body had been removed from the wreckage at around 4:00am. The body of William Palmer, of Toronto, was freed an hour or so later.

Working under the glare of huge portable lights, and using acetylene torches the "wrecking crew" from London toiled for more an hour before they succeeded in cutting through the mass to twisted steel to free the bodies.

Mayor William Anderson was among the early arrivals and one of the last to depart after the second body had been recovered. One of the injured men was removed to the hospital by taxi driver Albert King.

Firemen stood at the "ready" with lines of hose close to the point where railway rescue crew was working with their acetylene torches beneath the cab of one of the engines.

Missed Death

An unidentified motorist missed death by inches when he stepped on the gas and fled through the subway after seeing that a collision of the two trains was inevitable. In a state of collapse the motorist stopped his car in front of the home of Mrs. Arthus Crossman who invited him for a cup of coffee.

"He was shaking like a leaf," said Mrs. Crossman. "It never occurred to me to ask his name." "I'm going home," he said to Mrs. Crossman. "I've had enough for one night."

Throughout the long vigil, Mrs. Crossman proved to be a fairy godmother for shivering and weary newsmen. From 11 o'clock Wednesday night until 5 o'clock Thursday morning she was busily engaged making tea and coffee for the scribes.

Others Escaped

Although two were killed and four injured, several other members of the crews of the two trains escaped with a shaking up and minor bruises. They are Conductor J. Cowing, Engineer Tom Ready, Fireman J. Easby, Conductor G. Rivard, Trainmen Ken Shillington, B. Gale, all of London and Engineer Bert Hunter and Firemen A. Holden of Toronto.

The call for ambulances was received by Night Dispatcher Tom Nibel of Fraser's Taxi. Drivers of the taxis were Harold Despond and Amio Burkhardt.

Burkhardt was the first to reach the hospital with severely injured Henry Blocka, the Toronto fireman. 'When we arrived, he was lying on the sidewalk and evidently had been tossed a good 30 feet from the cab to the street. I could see he was in terrible pain but he kept urging the would-be rescuers to go after the others. "My buddy's in there, go get him-go get him." Blocka pleaded

Train Wreck Principals - Here are some members of the crews of the trains which collided in Galt the night of May 2. Front Row from left to right: Henry Blocka, Jack Easby, A. H. Olden, Herbert Hunter, Hartley Stewart, and the identity of the last person is unknown. Middle Row left to right: Gerry Rivard, Ronald Westworth, R. Gates, Jack Cowing the identity of the last two people are unknown. Back Row: Only the identity of the person on the far right is known. He is Tim Ready.

THE INQUEST

The following is a transcript of the Galt Reporter Article of May 17, 1956 on the Inquest of the May 2, 1956 Galt Rail Crash.

A combination of negligence and several other factors combined to cause the collision of a "slow" and "fast" freight train here the night of May 2, 1956, when two railroaders were killed.

This is the essence of a critical judgement rendered Wednesday afternoon by a jury probing the circumstances of the death of William Palmer, one of the victims. The other was youthful Tommy Watson of Galt. In addition to charging negligence on the part of the crew of the "extra" or "slow" freight, the jury under the chairmanship of Gordon S. Rouse, made certain recommendations designed to prevent a repetition.

Early in the hearing, Engineer William Hunter of Toronto testified that "if there is any responsibility for the accident, the responsibility is mine." The jury did not agree with Mr. Hunter in this view, its subsequent verdict revealed. The jury held that the responsibility could not be attributed to any individual, rather to "a combination of negligence on the part of the crew of the "extra" train, 5186, and lack of a proper system of communication with the station dispatcher."

Signals Inadequate

The jury also expressed the view that "the traffic blocking signals are not visible for all areas of the Galt Railway yards. "We therefore recommend, "the jury asserted in its written judgement, "a closer system of communication between the train crews, especially in connection with "extra" train crews and during unusual circumstances of the moving extra trains."

"We also recommend the use of a red flag by the dispatcher in circumstances similar to those that occurred at this accident, instead of taking a chance of someone not hearing, or misunderstanding an emergency order. We also recommend the installation of a better and closer blocking signal system within the Galt yard limits.

One of the highlights of the inquiry was the testimony of one of the witnesses that a split second before the crash, Engineer William Palmer realized the collision was imminent.

His Last Words

"Jump, jump" he cried to his fellows in the cab. These were his last words. The next instant he was dead. Fireman Henry Blocka of Toronto did have time to jump but, although severely injured, lived to tell the story at the inquest Wednesday.

Veteran dispatcher James Hartley Stewart provided one of the other highlights when he narrated his efforts to prevent the "extra" from pulling onto the main line as the "fast" through freight was nearing its destiny with death and destruction.

"Don't move, don't move," Mr. Stewart cried repeatedly, as he ran toward the crew of the "extra" waving his arms. But Mr. Stewart's words went unheeded, or unheard, for the "extra" pulled out and into the path of 903.

All witnesses were questioned by Crown- Attorney Harold Daufman of Kitchener, determined, apparently, to ferret out the real causes of the wreck. Once he demanded to know whether a witness would be able to offer a suggestion "which would give the public a guarantee the incident would not be repeated."

Both trains were drawn by "double headers" or, in other words, two locomotives of different types. It was also brought out that the crew of the "extra" train had received no orders with respect to Galt and that it was "on its own upon arrival here."

Not Informed

The fast freight was two and a half hours late in departing from Toronto and had lost an additional hour at Guelph.

Testimony also showed that although the crew of the slow freight knew the fast freight was behind them, they had no idea of how far, or when it would arrive in Galt.

The jury personnel were as follows: Angus A. Ferguson, Fred Bendus, David Brayshaw, Alexander (Sandy) Leith and Gordon S.Rouse, foreman. Detective-Sergeant Allan Slidders was the coroner's constable. Presiding Coroner was Dr. Charles Gordon Powers, of Preston

Still in the Hospital

Also critically injured in the collision was Engineer George Nutkins of London who, although he is in a cast at the South Waterloo Memorial Hospital is slowly recovering. He sustained multiple fractures to the arms and back.

Mortician Maurice O'Brian of Preston, who removed the bodies to the morgue, said he had recovered the watches and wallets of the victims. On checking the watches of Mr. Palmer and Mr. Watson, Maurice found a time difference of 55 seconds.

Photographer Glen Forbes identified the pictures which he had taken at the scene the night the collision occurred.

First member of either train crew to testify was Fireman Albert Blocka, a fireman on 903, the through freight. Still not fully recovered from injuries, he limped to the stand but declined a proffered chair. Mr. Blocka said that he had been employed by the CPR for four and a half years. The through freight of which he was a crew member, had priority over all trains. They had left Toronto at 6:35pm about two and a half hours beyond the normal starting time.

The two who died in the crash, Mr. Palmer and Mr. Watson were in the cab of the lead engine of 903.

Time to Jump

At the instant Mr. Blocka had caught sight of the other train, he had also heard Engineer Palmer's warning to "jump, jump". He jumped and alighted on the steps at the end of the platform and rolled to the street. Then, he regained his feet and managed to escape being crushed by the falling engines as they plunged into Water Street. He believed, he testified, Brakeman Watson was "giving a highball" with a lantern just before the impact.

Conductor Jack Cowing of London, who was in charge of the fast through freight said he was sitting in the caboose when the train came to a jolting stop, after prolonged application of the brakes. "When we reached Guelph junction" he said, "we were on time but we lost an hour there."

"Then you were hurrying to make up time," suggested Crown Attorney Harold Daufman. "No," the witness disagreed, "you can do no better than the time table schedule on the run." Long experience in railroading would enable him to estimate the speed of the run even while riding in the caboose. Mr. Cowing stated that the distance of 18 miles between Galt and Guelph Junction had been covered in the interim of 9:05 and 9:35pm.

After Engineer Hunter had shown a willingness to accept responsibility for the crash, Crown Attorney Daufman asked: "Why do you say that?" Because I did not go to the station operator to ask the whereabouts of 903," replied the engineer. Mr. Hunter then said it was customary railroad practice to rely on the brakeman to perform this duty, "but he the (brakeman) was not there."

Moved to Main Line

"I moved my engine onto the main line after it had been uncoupled," Mr. Hunter continued, but was interrupted by the Crown. "Why wouldn't the engineer of the second locomotive be responsible?" pressed Mr. Daufman. "Because he would assume that everything was okay," was the reply. "Surely railroads aren't operated on the basis of assumptions," suggested Mr. Daufman.

"We've had close shaves in the Galt yards before but nothing like this, " Dispatcher James Hartley Stewart told the coroner and jury. Mr. Stewart is a veteran railroader with 30 years of experience. "When the indicator went down in the office, I asked the operator whether it was 903, and he replied "yes". I ran to the siding and yelled to the crew of the extra: "Don't move, don't move."

"How close were you to the crew of 5186?" the Crown asked. "About 30 feet," Mr. Stewart said. When Conductor Cowing was asked how it is possible to check positions of various trains, the reply was by telephone. There were three such phones between Galt and Toronto he said. "Did you check on one of these phones?" the Crown asked. "No", replied Mr. Cowing, "the onus was on the inferior train (5186) to check." He added that the indicator in the station office would "show the train out of Killean" and the operator would know 903 was coming through because 5186 was already in the siding.

Poor Visibility

Fireman William Hopkins, who was on the second engine of 903, said he thought he saw something on the bridge, but the visibility at this point is poor because of the curvature of the tracks. "I hollered to my mate, George Nutkins to soak her but the air was gone. " This was attributed to the fact that the brakes had already been applied by Engineer Palmer on the first engine of 903.

Rear Trainman Ronald Westworth, London, agreed with his conductor that 903 had not made-up time between Guelph and Galt.

He added that the "order board is on the left but you have to be on the right side to see it because of the curvature of the tracks." Mr. Westworth stated that he did not realize there had been an accident until he had moved to the front of the train.

Picked up Bodily

At this point the witness injected an element of humour into the proceedings when he asserted, he had been taken to the hospital although not injured. “What was the matter with you?” asked the Crown.

“Nothing” witness replied, "someone picked me up bodily and placed me in the ambulance. Then it sped off to the hospital.”

At the request of the Crown, Timothy Ready, London, the engineer of the second engine on the slow freight, was granted protection by the court. Mr. Ready said the crew of his train had "no orders with respect to Galt." Instructions to drop off certain cars here "had been a message." He also said his crew would not know that 903 had been held up in Guelph.

"When you got to Galt then, you were on your own," suggested the Crown. "Yes," replied the witness. Answering further questions, he said there are block signals at the East and West ends of the yards. "We had to push off the lead engine onto the main line," he said. "We could see him moving."

Signal was Green

"There are no signals to control movements at this point," Mr. Ready said, "The signal for us was green." "How do you know when to move or stop?" pressed Mr. Daufman. "That's hard to answer. I don't know how to answer it," was the response after considerable reflection.

"What guarantee can we give to the public that this won't happen again?" Mr. Daufman continued. And what would you do to avoid a similar incident if you came through next week?" Witness did not reply. " I didn't know it was there," the witness said. "Did you look?" "No. Another thing - my vision was restricted."

Not Blaming Anyone

"What you are saying is that you were driving blind," commented Mr. Daufman. "Do you blame the brakeman?" "I won't say the brakeman was to blame," Mr. Ready said. "If he had got a red signal he would have known something was wrong."

"How did you get your instructions to drop off cars at Galt," Jury Foreman Rouse asked. "By message " was the answer.

Fireman Testifies

A. H. Holden, from Toronto, the youthful fireman on one of the "extra" engines said he had been employed by the railroad for two years. "Suppose your engineer had a heart attack, what would you do," asked the Crown. "Put on the brakes," Mr. Holden replied.

Recalled to the stand following Dispatcher Stewart's evidence, Mr. Holden said: "Yes, I did think it unusual when Mr. Stewart came running out and hollering." He was replying to a question by Mr. Daufman. "But" Mr. Holden added, "I thought he was saying that 903 was on our tail."

He had been a fireman since March 28. "How much training are you given?" asked the Crown. "Well, we had six trial runs," the witness said. Other witnesses agreed the "fast" freight had the "green" all the way through Galt.

R. Gates, London brakeman of the "slow" freight said 14 cars were dropped off at Galt. Before testifying he had been granted the protection of the court, at the request of the Crown.

He gave the signal to proceed to the engineers and presumed both had seen it. He said one of the engineers has handed him a "list " and said "cut me off." (Uncouple engines) "Then, I opened the switch and the signal to proceed. Then the engine moved onto the main line."

Saw Crash Occur

"I could see half way to Bond street and the Galt station, " the witness said. He added that he had seen the crash occur and two crew members flying through the air. Then the freight cars started to topple, so the witness had "taken off down the slope."

Fireman J. Easby stood mute when the Crown asked if he knew how this accident could happen?

After asking whether any of those present could throw any further light on the subject, the Crown suggested to the Coroner that the jury be permitted to retire. The inquiry lasted four hours. The jury brought in its verdict after deliberating about 45 minutes.

CANADIAN PACIFIC

LEN MILLS' STORY

Note: The following article originally appeared in the Cambridge Reporter on May 2, 1996.

I also interviewed Len in June 1999. Below is Lens version of that fateful night.

In May, 1996, my daughter Nancy, will be 40 years old. This brings back memories of around the time she was about to arrive in this world. I was 21 years old and had just started a new job on May 1, 1956 as a probationary firefighter with the Galt Fire Department.

I was assigned to the night shift which caused some concern to me because Nancy was to arrive at any time.

The first night (May 1) was uneventful with no fire calls. The second night (May 2) would be a much different story. The shift that night consisted of Capt. Fred Slee, Lt. Bob Murphy, firefighters Jack Plumridge, Ken Walker, Tom Mercer, Alex Nicoll and myself.

I remember feeling the building shake for a second around 10 p.m. Capt. Slee and I both wondered what it was. At 10:02 p.m., the second call of the night came in. A boxcar was on fire on the CPR bridge. To all of us, this meant the large bridge over the Grand River.

As we came to the top of the Water Street hill, the scene was something I will never forget as long as I live. The bridge over Water Street had collapsed and was down on the street. A large steam locomotive was lying on its side. Hot burning coals had spilled onto the street and a bright red glow was reflecting up on the large volumes of escaping steam from the ill-fated locomotive, producing what looked like a giant fire.

Over on the right side on the sidewalk, an injured crewman lay, calling for help. Fred Slee and I ran over to him. He said "leave me here and get out—that engine is going to blow up. " We knew we had to get him out of there fast. His injuries looked like he may have had a broken back. We needed to put him on a backboard.

I ran to the nearest house and ripped the door off their back shed. Don't ask me how I did it. but in a few seconds, I was back with the make-shift backboard and we carefully removed the injured trainman to safety. We kept telling ourselves the engine would not blow up as long as the steam was escaping in large volumes. Thinking back later, we were very lucky. At about that time, Fire Chief Walter Scott arrived on the scene and began piecing together what had happened.

A slow-moving train was pulling out of the station onto the main track. A second fast-moving train was proceeding west on the same main track. They collided on the Water Street bridge, collapsing the bridge and sending steam locomotives down each side of the east bridge embankment. This meant there were two more locomotives on the other side of the wreckage that we had not seen yet.

Chief Scott asked me to get over to the other side (north) and give him a report. It meant climbing up and over the west embankment and down the other side. As I proceeded towards the embankment someone in the crowd called to me that he had witnessed an automobile proceeding north on Water Street just as the bridge collapsed. He said, " I don’t think the car made it through in time."

As I arrived on the north side of the wreck, I was met with another scene right out of a war movie. Two large locomotives were lying on their sides, one on the road and one down the embankment on Samuelson Street. Equipped with a hand lantern, I climbed up the side of the second engine, knowing what I would probably find inside. No one could survive such a terrible accident. My worst suspicions proved correct. Inside the cab of the engine was a scene imprinted in my mind for life. Two trainmen lay dead.

As I began my descent back down the side of the engine, there stood another member of the train crew who was about to start climbing up. I said, "You cannot come up here." He stated, " I have to get in there and get my buddy out." I then told him everyone in there was dead.

"There is nothing you can do now and I can't let you in there." He did not comprehend anything I said and kept coming toward me, saying over and over, "I have to get my buddy out.'

I knew I had to do whatever I could to stop this determined young man from seeing what I had just seen. I tried pushing him back and we both fell to the street below. Again, he started for the cab of the engine. He seemed to have extraordinary strength. This time, he was not getting by me, I hit him with everything I had. As his knees started to buckle, I slid my arms under his to hold him upright.

Over to my left I saw a police officer (Lloyd Bender) and called out to him for help. Lloyd and another officer rushed over and took charge of the young trainman. "What happened to him? " Lloyd asked. "I don't know," I said. "He seems very weak." "We will take care of him—he will be OK in a few minutes." "Thanks," I said.

We learned later that one of the dead crew was a young man named Watson. I had known him some years before at GCI and VS. This was also his first night on the job. His father (also a railroader) lived across the Grand River and as the train crossed the CPR bridge, young Mr. Watson was to "highball" his dad.

CPR rules stated Mr. Watson was to leave the engine at Guelph Junction and proceed to the caboose to eat his lunch. He asked his replacement if he could wait until Ayr to switch and he agreed. Thus, Mr. Watson was killed instead of his replacement. This was the man I had just hit.

As I made my way back to the south side to make my gruesome report to Chief Scott, it suddenly dawned on me what had just happened. I had just struck a private citizen and a member of the train crew. I became very scared. I was there on probation, on the job for just two days. The man was sure to complain to the police who were escorting him away from the accident scene. All I could think of was, I will be fired in the morning.

Once over the embankment and down the other side I came face to face with Ken Walker, "What's wrong with you?" He said, "You're as white as a ghost." "I'm okay. I just found two train crew dead, but I am really worried about the baby coming. They tell me the rear of the train is blocking Dundas Street and with Water Street closed as well, how could get a taxi to the hospital?"

Ken just smiled and said, "Don't worry Len, they will go by Beverly Street and get around the end of the train on Samuelson." "Of course," I said. "I'm worried for nothing." I did not tell him my real worry, that I would probably be unemployed in the morning. We spent the remainder of the night searching for other victims and the car that was supposed to be under the downed bridge and wreckage. Luckily no vehicle was found.

No one ever mentioned the trainman I had "socked." In all the reports that followed, not a word was ever printed as I watched the Galt Reporter each day for a story on how a trainman got "slugged" trying to reach his buddy. That is to say, until a couple of years later when a strange event happened.

My neighbor had invited me over one Christmas Eve for a glass of cheer. I was introduced to his son, Ronald Westworth who was visiting from Lucan. He was a railroader. We chatted about railroading for a while, then he asked me what I did for a living. "Oh, I am a firefighter with the Galt Fire Department." A strange look came over his face. He said, "There is a guy in your department and if I ever meet up with him, I am going to knock his block off."

'Why, what happened?" I asked. ' I was involved with a train wreck a couple of years ago and one of the firefighters hit me when I tried to get into the wreck to find my buddy. I had better not ever meet up with this guy."

"Well," I said, "I'm your man. Take your best shot. But first let me tell you, I had to stop you from going in there at all costs and if I had to do it all over again, I would do the same thing again." Surprisingly, he smiled and said: "Yeah, I know. I was right out of it. I did not appreciate it at the time, but I guess you had to stop me somehow." We shook hands and that was the end of it.

I left the Galt Fire Department in September 1972, and went to Arnprior as fire chief, then on to Barrie as deputy chief. In 1979, I joined the office of the Ontario Fire Marshal, retiring Jan. 31. 1995. with the rank of regional manager. To end on a happy note, Nancy arrived on May 5, 1956, with no problems. She was a beautiful, very happy baby with not a worry in the world, especially a train wreck on the way to a hospital.

LOCOMOTIVE WHISTLE SALUTE FOR YOUNG VICTIM IN GALT CRASH - RAIL BRIDGE BACK IN SERVICE

The following is an article by Record Staff of The Evening Reporters published May 3rd, 1956.

A young Galt brakeman was buried today, but a train crew paid its tribute last night-in a long mournful blast of a locomotive whistle

The simple gesture - known as the "cannonball" tribute - was made as a special CPR engine, cleaned and polished, ground slowly past the white house that had been the home of 19-year-old Thomas A. Watson.

The youth was killed in a two-death train pileup that wrecked the Galt overpass 48 hours before.

First Across the Bridge

The engine saluting the youth last night was the first to pass over a temporary wooden overpass. The Watson home is nearby at the Highway 97 crossing. "It was the traditional railway way of saying hello and goodbye to Tommy," his mother, Mrs. A. K. Watson said today.

The 19-year-old lad had been making his first paid run as a brakeman on freight No.903 when it was involved in the accident Wednesday night.

"When Tommy was learning the job, every engineer with whom he rode would signal the house with two or three sharp short whistle blasts as passed the house." said Mrs. Watson. "We would flick the lights on and off to let him know we caught his signal. Then Tommy would swing a highball with his lantern."

Many Cards Received

She said the family has received cards and letters of sympathy from railroaders all over Ontario expressing sympathy and regret that life of a prospective railroader had been cut off almost before his career began.

The funeral service this afternoon at Little's Funeral Home drew a large attendance.

William Palmer, the other man killed in the wreck, was the engineer on the train. It was reported he was the engineer of the CPR Dayliner which collided with a cement truck near Dixie March 17, 1954. In that accident Albert Gillespie of Ayr was killed. Mr. Palmer spent nine months in the hospital.

Many of the 100 men from the railway had been cleaning up the debris and building the bridge for a straight 48 hours to finish the job. The first passenger train over the wooden structure was a 15-car train.

Bridge Unobtainable

It may be some time before the railway can secure steel for a replacement steel span. Railway officials were unsuccessful in obtaining an 81-foot bridge span from eastern Canada.

The wrecking crew have made provision for two lanes of traffic under the overpass. Water Street handles heavy traffic as it is part of No. 24 highway.

CPR officials investigating the cause of the collision of the doubleheader freight with the westbound extra have not made public a report on the cause. More information will likely be revealed at an inquest May 16 at 2pm at the Galt City Hall. Inquests into railway fatalities are compulsory under the Railway Act.

THE WATSON FAMILY

The death of 21-year-old Tom Watson, on his first paid run as a brakeman was tragic. His father Alex and mother Hanna were waiting outside their home, located near the CPR tracks, on 160 Blenheim Rd. waiting to waive to their son. Fate played a cruel hand and within seconds of passing by his parents the rail crash at the Galt station ended his short life.

His last words when he left his house was, "S'long pop, I'll give her the old highball when I pass the house."

Tom Watson, was born in Toronto and had worked at the Galt station in the baggage room and as a spare brakeman for two years before the accident.

Alex Watson, born in Scotland, moved to Canada in 1911. He met Hanna his wife in Toronto and worked for the Canadian Pacific Railroad for the railroad for 35 years, the last 12 years in Galt in signal maintenance. He worked out of the Dumfries CP station. He bought the station and moved it to 160 Blenheim Rd. as the first family home on that site. The station was replaced by another home and finally a large white home stood at the site for a number of years. Eventually the Cambridge Fire Department burned down the abandoned home.

All who knew Alex Watson say that he never fully recovered mentally from the loss of his son. Tragedy struck the Watson Family again when on Saturday morning, December 7, 1963 Alex allegedly committed suicide by walking in front of a CPR Dayliner train heading to Toronto south of the Roseville crossing near the Galt city limits. The Watson family home was adjacent to where Alex was killed. No inquest was held.

Hanna Watson was able to overcome the tragedy in her life and remained a part of the community until she died in January 1998.

Paul Watson was the youngest child of the family and did extremely well in school. He topped his class at University of Toronto for four years in Honours Arts and Archaeology. He then went to Yale University on a fellowship. He has for a number of years taught at a university in the United States. His memories remain with him of the accidents that took the life of his father and brother.

On Tom and Alex Watson's tombstone it is written, "Safe in the Arms of Jesus."

RONALD WESTWORTH INTERVIEW (1998)

Ronald Westworth - 1954 Galt Station, 1998 Elliot Lake credit: Ronald Westworth

Ronald was the brakeman for the double header through freight train #903. He was supposed to change with Tom Watson before Galt station but Tom wanted to stay up front so he could "highball" his parents who lived just past the station. Ron agreed and stayed in the caboose. Ronald was not injured in the ensuing crash.

At the time of the interview in 1999, Ronald had retired from CP Rail and lived with his wife Margaret in Elliot Lake, Ontario.

Ronald, can you give me a brief work history?

I started working for the Grand River Railway in 1954, then owned by CP Rail. They reduced the passenger trains at that time and we were transferred to London in 1955. I worked for CP Rail until 1963 when I quit because they were reducing the work force and I was only working spare. I drove City of London bus till 1965 then started working with Canadian National Railway. I worked at CN till 1969. I rejoined CP in 1971 in Windsor and worked for them in various capacities and cities up till my retirement in 1997. I also had a brief stint with GO Transit in 1995.

What type of impact did the crash have on your life and that of your co-workers?

Back in those days they did not have any stress or emergency trauma counselling. You just went back to work. It would have helped the guys if there was some type of psychological counselling dealing with the stress. Fireman Jack Easby quit right after the crash. He was in extreme shock. Tim Ready was also in shock. Toms father Alex Watson was never the same after and committed suicide in 1963. Many of the crew had a hard time coping with the accident. I myself was nervous and edgy for a few years after the accident.

You had a chance to read the rail crash inquest information. Were there any statements made at the inquest that you felt were not answered adequately?

Yes, one answer that was given by a crew member was that there were no orders to work around in Galt. Well, you do not need any orders to work around Galt. You are given your orders when you leave your base and it tells you what cars are being dropped off or picked up at what stops. You do not need any orders to work in Galt.

The train #5186 in the siding needed permission from the dispatcher to leave the siding when the superior train (903) was overdue.

Another statement was made regarding how you contact other trains. It was stated that there were three phones between Toronto and Galt. That is simply not true. You can stop anywhere and hook your line to the wires. But you are not going to be talking to other trains. You are going to be talking to the train dispatcher.

Was there any information that did not come out in the inquest that should have?

Killean used to be at the top of the hill, it was about five miles east of Galt. When our train went through at Killean, it would have triggered an electronic circuit that would have turned on a small indicator light in the Galt station. This would tell the operator that the train was coming out of Killean and he could ensure no trains would move out of the siding at Galt.

The electronic indicator would also be triggered in a small silver box on the siding switch. The indicator was down. The person throwing the switch, if he checked it, would have known if the train was coming on the main track or not. Nowadays the switch is automatically locked when the indicator is on. The switch cannot be opened.

The station operator, Stewart, knew that train was coming for at least twenty minutes and failed to notify the train crew in the siding. The engineer William Hunter acknowledged that if anyone was to blame it was him. He was right. Hunter should have gone into the station to ask permission to go on the mainline. He also gave the order to the Trainman Gates to open the switch. It was unclear if Gates knew the indicator position was down on the switch. It is the responsibility of everyone on the crew, not just the engineer, to ensure the train rules are followed.

Are there any other interesting stories from that tragic night?

A reporter called my pregnant wife at one in the morning the night of the crash to say I was dead and did she have a picture of me. I was actually fine but, in the hospital, briefly.

WILLIAM HOPKINS INTERVIEW (2002)

William Hopkins Interview 2002 credit: author collection

William "Bill" Hopkins was the fireman on the second locomotive #5420 of "through" freight #903.

1948

On January 2, 1948, I started in the railway working out of John Street in Toronto, doing my shop time. I started out as a labourer and then as a machinist's helper. They wanted me to stay in the shops but I wanted to go out and "play" trains. The money was better working outside. I was a young man then and I wanted to travel around.

You either went head end or tail end at that time. I started as a fireman, afterwards I was classified as an engineer. I did some spare runs. You just gravitated through the seniority list.

May 2nd, 1956

What went on that day was nothing unusual. It was a miscalculation that is all. The only saving grace, as far as I am concerned personally, was that Bill Palmer (lead locomotive engineer on the "through" freight) was going slow. It was his first trip back after fracturing his skull, when the train he engineers on hit a cement truck in west Toronto. He was off for quite a while. He came back and this was his first trip west so he was going slow.

We were going through Galt, and we had lots of reason to fly through there. We used to violate every rule in the book at that time relating to speed going through Galt. I believe it was 10mph through Galt. We never did 10 mph. Bill did go 10mph. I would suggest, had he not gone the speed limit, that there would have been no head end crew left, no station left and the first two houses on Water Street would have been gone also. Usually, once you get a clear order board, we would go full out, stick it to her. I know guys who have gone through there at 50 to 60mph.

We all knew what happened. I went to the inquest. It really hurt William Hunter (engineer on lead engine of slow or "extra train that came out of the siding that was hit by "through" freight #903) because Bill Palmer, who was killed, was his buddy. Once it started, he broke up, I asked if we could be excused. I tried to help him. We went across the street to the hotel and had a drink to try and help Hunter get through the inquest.

I was off for about 18 days. You get thrown around in that cab pretty good. I was banged up in the upper corner but I was able to get out.

George Nutkins went out the window behind me. I had a little trouble convincing the police chief that there was another person still in the crash. I said there is a kid named Tommy Watson. They found Tommy and I just fell apart.

I was sitting in the station and the assistant superintendent was smoking a cigar. He said, "What are you doing here?" I said I was waiting for transportation home. He said, "You don't have to wait in here, you can wait outside."

The general superintendent Johnson was there. He had one main job in life and it was to get that line open. Jack Cowing, conductor, told the general superintendent who I was and the superintendent said, "Get this person to the hospital now!" Johnson called a cab and got me outta there. It was very sad.

We were way late in single freight #903. We sat an hour at Guelph Junction waiting for engine #2205 to come. We were late because we were supposed to be diesels but they were not functional, so we had steam instead of the diesels. So, the guys in the train in the siding at Galt thought that #903 won't be on that time. So, it just so happens, we almost were.

What they were gonna do was get out of the siding, get up, get back in, so they could do what they wanna do around the interchange. It was just out and back in. Slam the switch and that was it. Palmer could not see ahead, that the "extra' train had moved out of the siding, there was a curve. George (Nutkins) had dumped the air. Once the air goes you know something is wrong. Anyway, that is all there was to it.

William Hunter, Jack Easby and Tim Ready were all suspended for a year as the result of that accident. I don't know if Ken Shillington, brakeman on the second locomotive in siding, ever came back. He was a sharp young man.

1957-58

The whole culture was changing. I became local chairman of the Fireman Union, Chief Griever. Not long after that we had a nine-day lockout in January. Then we had a strike in the summer. At that time the total number of firemen on the CPR was 2,855. The CPR on the first strike was only able to run passenger trains - no freights.

I went to Ottawa to speak at a Royal Commission. The policy attitude of the CPR was that they were the front man for the American railroads. The Canadian union was being supported by the American Union. The CPR never got rid of the Fireman when I was there. It was a cut and dry deal though. At the time they were putting in the Diesels, I couldn't see the reason for the Fireman.

1960

The prospect of spending three or four years of being a hogger in Windsor, a couple years on the spare board and spending years chugging up and down that line with diesels till 1 was 65 years old did nothing for me. I decided to leave the CPR.

Steam engines were fun. Sure, they were dirty. If they still had the steam engines I would probably still be with the CPR. All the time I was working with the steam engines, no one ever wanted to ride up front. When Diesel came everyone except the secretary wanted to ride up front. I was still a young man when I wanted to get out.

Later I worked in sales and then ten years with the provincial government.

CANADIAN PACIFIC
SERVICE
414320

GEORGE NUTKINS INTERVIEW (2006)

The following is an article by April Kemick that was published in the London Free Press on May 1st, 2006.

The horrific sound of grinding metal and the earth-shattering crash that followed will haunt George Nutkins forever.

The 82-year-old Londoner defied death that night half a century ago tomorrow, when the steam locomotive he was travelling in plowed into another train in Galt, now part of Cambridge, killing two co-workers.

It was May 2, 1956, the last big wreck of the steam-train era for the Canadian Pacific Railway -- and Nutkins says it shouldn't have happened.

"That train wasn't supposed to be there," said Nutkins, then an engineer for the CPR. "We had the green light."

As Nutkins and his crew -- many of them from London -- travelled west through Galt en route to London, they could see a train sitting parallel to theirs on a different set of tracks.

With the green light to proceed, Train 903 -- Nutkins' unit -- plowed ahead as scheduled. In a split second, everything changed.

"We could see that the train had pulled out of the siding," said Nutkins, who was riding in the second of two engine cabs with a co-worker when the trains collided on a bridge.

"I didn't have time to do anything," he recalled. "I really didn't expect to get out of it alive."

Instinctively, and with only seconds to spare, Nutkins sounded the train whistle to warn any pedestrians or drivers of the impending crash.

He hollered at his co-worker to jump, said a hasty prayer and braced for the impact.

"It was a terrible, grinding crash," said Nutkins, who, by stroke of luck, was thrown out a small window, landing more than 50 feet away on the street below the railway bridge.

If he hadn't been thrown clear, he would have faced certain death, either from the impact of the collision or by burning to death when the engine's steam pipes burst.

"It was an awful, awful thing." The scene, as a local newspaper described it, was “total devastation."

The legendary crash even inspired a book, Paul Langan's Tragedy in Galt.

Train 903's two steam engines, crumpled in the crash, straddled the sunken bridge as wrecked train cars tumbled down the embankment. Coal was scattered all over the road and downed power wires were strewn across the bridge.

His battered body on the street below, Nutkins slipped in and out of consciousness. But while he suffered 21 fractures and spent nearly two years recovering in hospital, Nutkins knows he was lucky.

Unlike engineer Bill Palmer and young brakeman Tom Watson, Nutkins survived to finish his career on the rails, eventually retiring as superintendent of the London division.

"Two people were killed," he said, tears welling in his eyes. "I think God was on my side that day."

KEN SHILLINGTON MEMORIES (2020)

The following article is a summary, of emails received and listening to Ken talk about the crash, by the author in 2019-20.

I received an email in 2017 from Steve Shillington who said his dad was interested in getting a copy of the 1st edition of my booklet *Tragedy in Galt*. Imagine my surprise when I realized that his father was Ken Shillington the brakeman on the second locomotive of the "extra" Train #5186 from the 1956 crash.

I had mistakenly thought all former crewmen from the accident had passed on. Two years later Ken came to Cambridge to talk about working at the CPR Railway and the crash.

Ken was 24 years old and had a young family when he started at the Canadian Pacific. He was on his first paid run as a brakeman that fateful night of the Galt crash.

Ken's recollection before and after the wreck were fine. Unfortunately, Ken suffered from Post-Traumatic Stress due to the accident and can't remember anything about the actual accident. He remembers walking away to see the wrecked engines on Water Street and the escaping steam but little else.

He did remember the conductor, Gerry Rivard, was worried about keeping his job as they ran to Water Street to view the wreck.

The next thing he remembers is that he had to work a trip to Orr Lake in the middle of the night. During the trip to Orr Lake, he was positioned up at the hand brake on the top of the last boxcar in case a connecting knuckle had been broken in the crash.

The first part of the trip to Orr Lake was very stressful as it was uphill and if the knuckle had been broken, the boxcar would have rolled down to the site of the wreck. Standing on the top he had just passed the wreck before the train headed uphill. It was scary.

When he arrived home in London, he drove himself home. No one from CP spoke to him about what had happened the night before. He was not asked to attend the inquest.

Ron Westworth, head end brakeman and crash survivor did many trips with Ken after the crash and they never discussed it.

Ken continued to work for the Canadian Pacific for two more years as a brakeman. Being on call at all times including weekends and hoping for more hours was stressful. He decided to look for more permanent employment.

REFERENCES

The Accident

(1956-05-03) Two Dead in Railway Crash, The Evening Reporter

Eyewitnesses Tell Grim Story

Kewley, C. (1956-05-03) Eyewitnesses Tell Grim Story of Crash and Rescue Efforts, The Evening Reporter

The Inquest

Kewley, C. (1956-05-17) Jury Finds Lack of Proper Communication System, The Evening Reporter

Lens Mills Interview

Mills, L. (1996-05-02) The Great Train Wreck, Cambridge Reporter

Locomotive Whistle Salute for Young Victim in Galt Crash - Rail Bridge Back in Service

Record Staff (1956-05-05) Locomotive Whistle Salute for Young Victim in Galt Crash - Rail Bridge Back in Service - Kitchener-Waterloo Record

The Watson Family

(1963-12-03) Alex Watson Death Notification, The Evening Reporter

George Nutkins Interview

Kemick, A. (2006-05-01) 50-Year-Old Crash Still Vivid – London Free Press

PHOTO CREDITS

Front Cover

Kitchener-Waterloo Record Photographic Negative Collection, Dana Porter Library, University of Waterloo

ISBN Page

Crowd looking at crash - Courtesy of the City of Cambridge Archives

Introduction

Crash photo - Kitchener-Waterloo Record Photographic Negative Collection, Dana Porter Library, University of Waterloo

The Accident

Train photo from above - Brian Sinclair

Eyewitness Tells Grim Story

Crowd scene - Kitchener-Waterloo Record Photographic Negative Collection, Dana Porter Library, University of Waterloo

Fire chief looking at wreck - Kitchener-Waterloo Record Photographic Negative Collection, Dana Porter Library, University of Waterloo

The Inquest

Train crews at inquest - Galt Reporter May 17, 1956

Lens Mills Story

Crash scene with men around mailbox - Kitchener-Waterloo Record Photographic Negative Collection, Dana Porter Library, University of Waterloo

Len Mills - Authors Collection

Locomotive Whistle Salute for Young Victim in Galt Crash - Rail Bridge Back in Service

Tommy Watson - Author's collection, original owner retains rights

The Watson Family

Tommy and Alex Watson - Kitchener-Waterloo Record Photographic Negative Collection, Dana Porter Library, University of Waterloo

Ronald Westworth Interview

Crowd scene - Courtesy of the City of Cambridge Archives

Ronald Westworth - 1954 Galt Station, 1998 Elliot Lake credit: Ronald Westworth

William Hopkins Interview

Two engines on the road - Kitchener-Waterloo Record Photographic Negative Collection, Dana Porter Library, University of Waterloo

George Nutkins Interview

View from above crash site - Raymond Bloye Collection

George Nutkins 2006 – London Free Press

Ken Shillington Memories

Two cranes at crash site - Kitchener-Waterloo Record Photographic Negative Collection, Dana Porter Library, University of Waterloo

Ken Shillington - Ken Shillington

References

Crane lifting engine on bridge - Courtesy of the City of Cambridge Archives

Raymond Bloye working on engine - Raymond Bloye collection

Photo Credits

Two engines on collapsed bridge - Kitchener-Waterloo Record Photographic Negative Collection, Dana Porter Library, University of Waterloo

Made in the USA
Monee, IL
04 November 2026

67375826R00036